Space Flight

by Janet Adams

I'm Jenna Walters and I've been a Private Investigator for the past several years, so I'm a questioning type of person. I have my own investigation agency here in Miami and have worked on a few crime cases in the last couple of years; some even made the news.

An unusual message was left on my answering machine yesterday. It seems there's a private space travel company located up the coast near Cape Canaveral. They take people on flights into space and back. Now, I would call that pretty unusual, wouldn't you? My interest is peaked.

Well, I returned their call and this is what I found out. "I'm Phillip Gardner. Two friends and I are starting a business called Space Flight. It involves building a space ship which will carry people into space and back to earth. We are using our own capital along with some borrowed money, to purchase the parts to build it."

"Why have you called me?"

"We talked with several people at Clayborn College, located in south Florida, asking if they knew of a graduate with an interest and an aptitude in science and a possible interest in space travel. They mentioned you, remembering how you were always questioning things and liked to read articles on the latest scientific research. Our being situated near Cape Canaveral is no coincidence. We plan to rent NASA's facilities as a location for training and to lift off and return from our trips."

"How do you see me fitting into all of this?"

"We need someone to check out prospective employees and travelers and to join the travelers as they train for their trip into space."

"In addition to your interest in the sciences, we've heard of your excellent reputation for knowing people and investigating their background. This trip is not without danger and we want to eliminate taking anyone with us who wants an early and spectacular demise, for whatever reason. So, for sure, we need to check them inside and out by every means possible. They will be signing a mired of release forms to facilitate this process and, they will also be paying for their ticket. Would you be willing to meet with my associates and myself over lunch so we can tell you more about our adventure?"

"Yes, I'd like that. Tell me when and where."

They picked a seafood restaurant halfway between Miami and the Cape. I met them there at 11:00am the next day. We introduced ourselves and followed the hostess to our table.

My phone contact was Phillip Gardner and his partners were Alex Wheeler and Robert Thomas.

"Ms. Walters, we'd like you to be a part of our company. How much do you know about space travel today?" Phillip asked.

"Mainly, what I see on the news or what I read in scientific journals. I know NASA hires privately-owned companies to carry supplies to and from the International Space Station."

Phillip continued, "Well, more has been happening in the space travel field than that. For instance, it took more than 35 years and a journey over 15 billion miles but Voyager I was the first man-made object to make it into interstellar space.

"NASA scientists are sure of this much: first, astronomers found, through Voyager I, a steep drop in solar wind coming from our sun and secondly, they noticed an up-tick in galactic cosmic rays coming from outside our solar system. These two events suggested that Voyager I had crossed the edge of the heliosheath. The heliosheath is a boundary marking where these two winds meet and form a bubble around our solar system."

"But I digress because I see your eyes are glazing over. We won't be going that far in our flights. I just wanted to give you a little idea of how far humans have come, even with some technology from 35 years ago."

"Well, I'm kind of glad you won't be going that far. That's a lot to wrap my mind around. But going back to your other question about my going with your company, if I went with Space Flight, would that mean I would have to move from Miami to up near the Cape?"

"Yes. We've scouted out the area near where our headquarters are located and have several places you can choose from. We'll pay all your moving expenses, the cost of your rent, food, etc., a handsome salary plus a car, gas, and an expense account. We'll be working closely together with you in all personnel areas, both in who we hire for the company and also for potential customers / travelers. How would you like the title of Vice President for Personnel?"

"That sounds great. But I really need some time to think about all of this. Could you give me some reading material about your company so I can learn a little bit more about its leadership, and goals?"

"Yes, here are some brochures with photos of our headquarters and the information you requested about our company. Would it be alright if we contact you in a week?"

"That's fine," shaking hands, "I look forward to hearing from you in a week."

The ride back to Miami was a quiet one. So much potential change was in the offing, although Cape Canaveral is only a 5-hour drive from Miami. My world, despite my being a P.I., has been on the quiet side the past few years. I sense that's about to change. One thing that will remain the same is my relationship with God. Can I adjust to the fact that my God is far bigger than I thought; that He's not just the creator of earth and all its life, but of the universe and beyond? I can see this will take some time to percolate inside me.

Chapter 2

When I returned home, I unpacked, sat down and began reading the brochures from Space Flight.

I found that Phillip Gardner has a Doctorate in Aeronautical Science from an elite, Ivy League school. He interned with NASA and then worked for them for several years.

When he has the time, he likes to fish and read scientific journals.

Once a month, he is also involved in a ministry through his church, of feeding hungry people. He says this is a high lite of his month.

Talking to herself, "This sounds like a really nice, balanced guy, someone I would love to work for. His partners are equally impressive."

"If I went with Space Flight, I could rent my house out for a year and that would give me an excuse to go back to visit with friends whenever I have a break in my schedule. And the same goes for my office; I'll rent it to another P.I. for a year.

"I'm going to make an appointment for tomorrow to talk with one of my professors at Clayborn College, Dr. Thorensen, and see what he has to say about my going with Space Flight.

Chapter 3

Entering the university's familiar halls kindled many good memories. Dr. Thorenson was one of my best. I always appreciated his counsel. Here's his office. Knocking, the door opens. "Dr. Thorenson, how nice to see you," extending her hand. "It's been a few years."

"Yes, it has," shaking hers. "To say I was surprised when I received your call, would be an understatement. Please come in and sit down. What brings you here?"

"I've been offered a job as Vice President for Personnel for a company located near Cape Canaveral. It's called Space Flight. It specializes in flying people into space where they experience weightlessness, participate in some experiments, spend one or two nights there and return to earth."

"Who finances this grand endeavor and to what end?" asked the professor.

"The private citizens who go will be paying their own way. Also, several cameras will be mounted throughout the ship by a reality TV show company which will be paying for the rights to record the experience and broadcast it at a later date. We hope this will be the first of many trips into space with one day, traveling to the moon and beyond."

"Wow!! That sounds like the future is now. What part would you play in all this?"

"I would be in charge of checking out prospective employees for the company and also, for interviewing and thoroughly vetting all those who apply to be a passenger on a flight."

"It sounds exciting and challenging and a whole bunch of other adjectives that come to mind.'

"But professor, should I take the job?"

"Are you ready for an exciting and challenging new chapter in your life?"

"Yes."

"Then go for it and don't look back!"

Chapter 4

The week passed quickly, full of activities like contacting a realtor to handle renting my house and meeting with friends to tell them I was moving to Cape Canaveral but would be back from time-to-time.

The day came for me to hear back regarding the job offer. A little after 10:00am, the phone rang and it was Phillip Gardner from Space Flight.

"Well, Ms. Walters, do you have good news for me? Will you accept our job offer?"

"Yes, I will. I'm looking forward to being a part of your company."

"Great! I'll send a van and driver to pick you up and bring you here. Will a van be big enough?"

"Yes, I'm renting my house out fully furnished, so I'll just have a few suitcases and some boxes. Also, can I use Space Flight's address to forward my mail to until I get settled somewhere?"

"Absolutely. Will you be ready to come on Wednesday around 9:00am?"

"Yes, I will."

"Good. We'll get a hotel room here for you for a few days until you find a place. The van driver who is coming for you will give you the business card of a realtor we've used in the past. I've told her you'd be calling to make an appointment to go house hunting. By the way, I'll also give your driver a credit card in an envelope. It gets replenished with $5,000 at the first of each month.

"That sounds more than generous! Thank you

Chapter 5

After arriving at my room in the hotel, I laid down for a few minutes, then called the realtor. She had been waiting for my call and we made an appointment to meet the next day at 10:00am in the hotel lobby to go house-hunting.

It didn't take long to find a house I really liked. I signed some papers and headed back to my hotel. As I was walking toward the elevator, the manager called out to me, "Oh, Ms. Walters, you have a message." He handed me a small white envelope. Turning around and walking away, I opened it and began to read, "You will hire who we tell you to hire. If you value your life, you won't tell anyone about this note. Allah is great!"

I could feel my temperature rise and knew the color in my face was evaporating into a whiteness like the envelope I was holding. As I ascended the elevator, I questioned myself, "Who knows about my new position at Space Flight? And why would anyone care who I hired?" This is beginning to sound like the P.I. job I left in Miami. When I get to my room, I'd better check it for any listening devises or cameras. As my Daddy used to say, 'E Pluribus Unum.' Don't trust anyone....or something to that effect.

The room search was negative. I'm calling Phillip on my cell phone and see what he thinks I should do.

"Hello Phillip, this is Jenna Walters."

"Hey, did you get registered at the hotel? And what about the realtor?"

"Yes, yes, that's fine but, something came up that I need to talk with you about. No, I can't talk over the phone. Can you meet me in the hotel lobby in a half an hour? Alright, I'll see you then."

A half hour later, "Thank you for coming. Let's take a seat over here," motioning to a couch.

"So now, tell me what you couldn't tell me on the phone."

"Phillip, this note was left in my box at the hotel front desk," handing it to him.

After reading it, "We were warned by the federal government when we went into business that we might receive threats." Taking out his cell phone, "I'm going to call the F.B.I. immediately and tell them about this." Moments later, "I told them we'd meet them in the hotel lobby in about 20 minutes. They said not to use the hotel phone, only your cell phone."

"Phillip, I've only been here a little more than a day, how could anyone know where I'm staying, where I'm working, and what my job is?"

"Have you filled out any forms here at the hotel…maybe when you registered?"

"Come to think of it, yes, I did fill out some paperwork when I checked in."

"We'll tell that to the agents when they get here. They might want to talk somewhere other than this hotel. Here come two fellows now. Hello," extending his hand, "I'm Phillip Gardner, President and CEO of Space Flight and this is my Vice President for Personnel, Jenna Walters. She is staying here at the hotel for a couple of days until she finds a house to rent."

"Hello, I'm Chad Withers and this is Sam Guttenberg, F.B.I. agents, based in our Cape Canaveral office."

"Gentlemen, we have reason to believe there is a risk if we stay here to talk. There's a donut shop about a half mile north of here on U.S. 1. Why don't we go there to talk?"

"Fine." Chad adds, "I'll have a valet bring my SUV up front and we can all go together."

After riding about five minutes Phillip says, "There it is, up on the left."

"I see it. I'll pull in the back and park."

All four walk around to the front of the diner, go in, and are seated. After ordering, Phillip shows them the note addressed to Jenna. A half hour passes and they begin to exit the diner. Inside the

SUV, one of the agents, half-turning to Jenna, says, "So it's clear, right? When we get to your hotel, you tell the clerk at the front desk that you received distressing news from home and have to check out right now. We'll be waiting for you in front of the hotel. We have room for your luggage and boxes in the back of our SUV."

"O.K., I'll try and be back in 20 minutes or so."

"Jenna, I prepaid for the room so don't worry about payment."

"Thanks, Phillip." And off she went.

About 20 minutes later, she came back with a bellboy following, pushing a rolling rack full of luggage and boxes which he packed into the back of the SUV.

"Alright," one of the agents said looking back at the passengers and luggage, "Are we all set?"

"Wait! I just thought of something. Why don't I follow you in my car so I don't have to come back here to get it?"

"Good idea, Phillip. Here's the address of the hotel," handing him a card, "in case we lose you." After a few seconds, turning his head to Jenna,

"Ms. Walters, from what you told us in the restaurant, this hotel is not a safe place for you to stay. We have a hotel in mind where some of our agents from out of town stay. We believe it's safe."

After about a half-mile drive, they meet Phillip in the hotel parking lot and all enter the hotel together. Jenna checks in and gets a room key. One of the agents asks Jenna, "Would you like us to accompany you to your room?"

"No, I'll be fine. Thanks for coming."

"Let me help you with your bags," Phillip volunteers.

"Thanks, it's been a long couple of days." After a few steps, "I have no idea what this was all about."

"The world has been changing for some time now, overtly and covertly, since 9/11 and even before then. With the note you received being signed, 'Allah is great,' it's pretty obvious this is related to terrorists with Islamic ties. The war they're fighting is not on an actual battle field but subtly, in the fabric of our everyday lives, and in our beliefs. People who feel disenfranchised or left out-of-the-loop are ripe for that kind of talk. It's a kind of brainwashing."

"H-m-m-m, I think you're right."

"And they're trying to make headway in the aerospace industry, just like they tried with you, one person at a time. I think you're safe now." Taking her hand, squeezing it softly and saying, "Don't worry. By the way, were you looking for a house in a gated community?"

"No."

"Then call the realtor again and tell her there's been a change. Take tomorrow morning off to do house hunting and I'll see you at my office at 1:00pm."

"Sounds good. I'll see you tomorrow at 1:00pm."

A call to the realtor confirmed she'd meet me in the hotel lobby at 10:00am tomorrow.

Chapter 6

I awoke to the sound of jet engines. In seconds, I realized my hands and feet were tied. One cloth covered my eyes and another, my mouth. I was being flown somewhere. What happened? How did I get here on a plane going who knows where? My mind was still a little fuzzy, but I remembered last night. Phillip was bringing my luggage inside my hotel room. As he left, he squeezed my hand and said, "Don't worry. I'll see you tomorrow at 1:00pm." Then I called the realtor regarding meeting the next morning.

After that, I don't remember anything. I wonder how long I've been here. With this blindfold, I can't even see out a window to find what time of day it is or even, if it's day or night. I rolled on the floor a little and bumped into what I assumed were seats, two of them. If there are two seats on one side of the plane and two on the other, then this isn't a big passenger plane....maybe a private plane with a limited range of flight without refueling. Where are they taking me and why?

"Ah, Ms. Walters, I see you are awake. You have been sleeping a long time. Let me take off your blindfold and mouth covering so we can talk."

"What about my hands and feet?"

"That will come in due time. First, I need to know that you know that any efforts to escape are futile. We have been flying over the Atlantic Ocean for several hours and we will be landing soon to refuel and replenish our food and drink."

"You speak with an accent but I don't recognize it as Middle Eastern."

"My accent is not Middle Eastern because I am Kazakhstani and I work on the Russian space effort at the Baikonur Cosmodrome in Kazakhstan."

"I don't even know your name."

"I am Boris."

"And you are taking me to Kazakhstan? Can I ask why you are taking me there?"

"We are going there to rebuild our space program."

"I'm thinking you have the wrong person. I don't know anything about space travel."

"But you do know someone who does know a lot about space travel, Phillip Gardner, the President and CEO of Space Flight."

"Phillip Gardner just hired me as head of his personnel department. I haven't even started to work there yet."

"We know all of this. And we also know that you both have gotten close. We saw how he squeezed you hand as he left your hotel room."

"So you were in my hotel room before we got there. You must have spies in every hotel near the Cape. What was this, 'Allah is great' all about?"

"It was a way to cover our tracks."

"So you are a terrorist, just not an Islamic terrorist."

"No, we do not consider ourselves terrorists. We are doing this for the Motherland, Russia."

I look at my watch, still on Florida time, and it's nearing 1:00pm.

Chapter 7

Phillip, looking at his watch, says to his secretary, Mary, "It's almost 1:00pm. If anything, I would think Jenna would be early, wouldn't you?"

"Yes, I would."

Taking the hotel business card out of his wallet, he gives it to his secretary. "Will you call the hotel and see if she's there?" A minute or so passes.

"They said there was, 'No Answer', when they tried calling her room."

"Maybe she's caught in traffic. I'll wait a little longer. She was supposed to meet with the realtor this morning. Mary, will you look in your rolodex and get the realtor on the phone, please."

"Yes, sir," reaching for the rolodex and then the phone, she dials. In a few seconds she says, "I'm calling from Phillip Gardner's office, would you hold?"

"Yes, this is Phillip Gardner. I'm wondering how your meeting with Jenna Walters went this morning?"

"Well, I was going to call you. She never showed up in the hotel lobby. I called her cell phone around 10:00am but there was no answer."

"That's strange. I called her hotel room a little after 1:20pm and there was no answer. I think I'll go over to the hotel and see what I can find out. Thank you."

About 30 minutes later, Phillip pulls up to the hotel and parks. After entering the hotel, he asks the front desk person, "I'm looking for someone registered at your hotel, Jenna Walters. Can you call her room for me?"

"Yes, sir." Moments later, "There's no answer. She must have stepped out. Would you like to leave her a message?"

"Not right now. Thank you." Phillip steps away from the front desk, pulls out his cell phone and dials.

"Hello, this the F.B.I., Cape Canaveral, how can I help you?"

"I'd like to speak with either Agent Withers or Guttenberg."

"This is Agent Chad Withers. Can I help you?"

"Yes, this is Phillip Gardner. I was with you and Agent Guttenberg last night. It seems my new employee, Jenna Walters, has disappeared. Would you both come over as soon as possible?"

"We'll be there in 20 to 30 minutes."

"Good. I'll be in the lobby."

Twenty minutes later both agents walk up, "Phillip Gardner," shaking hands, "It's good to see you again. Tell us about Ms. Walters. Where were you when you last saw her?"

"After you both left last night, I walked her up to her room, left her and her luggage inside the door and told her I'd see her the next day, today, at 1:00pm."

"Did you walk with her through the rooms to make sure they were empty?"

"No, I assumed if the F.B.I. used the hotel, I could trust it was safe."

"Let's go up and see what we can find out. I got a key from the front desk clerk." After a minute or so ride up the elevator, "Here we are." He walks a few steps to the door, unlocks it, and they enter the hotel room. It looks neat and tidy, almost un-lived-in. "Let's check out the bedroom," as they walk toward it. "She never stayed here last night."

"And no luggage. I brought her luggage in and put it down right here inside the door," pointing down at the floor. "And now it's gone," his voice rising. "She had an appointment downstairs in the lobby at 10:00am with a realtor, which she missed." Phillip begins pacing the floor.

"O.K. Let's sit down over here," as Chad motions toward the living room. "Let's remain calm and go over the leads we have. The note she got said, 'Allah is great,' which indicates Islamic terrorists."

Listening, Phillip sees something on a table in the living room and walks over to pick it up. "This looks like a matchbook cover, but I can't read what it says. It's in another language. Chad, what do you think it is?"

Taking it and looking at it closely, "It's no Middle Eastern language that I'm familiar with. I think it might be, no....it might be Russian or from a Russian satellite. I'll have to confirm this with my office. We need to check any flights which departed last night heading east. Let's start with any private flights. Phillip, we're headed to our office now where we can check all this out faster. We'll call you when we find out anything."

The next day Phillip grows more and more apprehensive about Jenna's whereabouts and safety. Finally, the phone rings; it's the F.B.I. "Mr. Gardner, Agent Withers here, we've done quite a bit of investigation and we don't think Ms. Walters was the target. The main target here was you. The terrorists grabbed Ms. Walters to use as bait to snare you. You are the one with all the knowledge of space flight. We've traced a possible flight that left Orlando last night, with a declared destination of Kazakhstan."

"What do you mean by 'declared destination'?"

"By U.S. law, all flights are required to file a flight plan containing all sorts of information, including destination. In that column, they wrote Kazakhstan. Whether that's true or not, we don't know."

Phillip gasps.

Chad continues, "We know that before the Soviet Union fell apart in 1991, their space ships took off from the Baikonur Cosmodrome in Kazakhstan. We believe that's where they are headed now."

"God, no. So, do you think that they think I can help them with their space program?"

"We think that's a pretty good guess."

"What happens now?"

"We wait…we wait for them to contact you. And I don't think we'll have to wait too long. When you hear from them, call us immediately."

"Will the U.S. government let me go there? And for how long?"

"Mr. Gardner, if Jenna Walters is important to you, you had better get your affairs here in order. You have a U.S. passport?"

"Yes."

"O.K. They will have to arrange for a visa for you to Kazakhstan, which I'm sure they've already done."

Phillip hangs up the phone with the F.B.I. He thinks for a minute or so and picks up the phone again to call his Vice President of Space Flight, Alex Wheeler.

"Alex, it's Phillip. I have some bad news. Some terrorists have kidnaped Jenna Walters. The F.B.I.

thinks they've taken her to Kazakhstan and are using her as bait to try and lure me to go there and work for them to get their space program going again. Alex, I can't say, for sure, what's going to happen but the F.B.I. told me to get my affairs here in order. Your signature is already on our business bank account and I want to put you as a signer on my personal account."

"Phillip?"

"Alex, I have to play with the cards I'm dealt. Until I know more, I'm placing you in charge of the company. I anticipate we'll be able to talk by phone from time-to-time. But I don't know how long all this will take. Kazakhstan time is 10-hours ahead of Florida."

"Phillip, you can count on me to run everything as we planned until you get back."

"Thanks, Alex.

Chapter 8

Boris, will you please untie these ropes on my arms and legs? They are getting cumbersome, to say the least. And tell me what this is all about."

"Ms. Walters," as he unties the ropes, "In May of 2015, a Russian Soyuz space craft, Progress 59, cargo mission to the International Space Station, fell back to earth without reaching the space station. We believe it was because of a problem in the third stage of the rocket. But we are not ruling out sabotage."

"Boris, I am really, very sorry this happened, but I don't see how I can help you. I know absolutely nothing about rockets."

"Ms. Walter, as I told you before, we know this. We want Mr. Gardner to come here and get you. I know he can help us. We want to make sure this third stage is fixed before we let any crew ride on top of it."

"Boris, does your government know you are doing this, or are you just kidnaping me on your own initiative?"

"My brother, Igor, and I have worked on the Russian effort to resurrect our space program for several years and we don't want to see it collapse because of one mistake, a mistake either he or I made."

"When you contact Phillip Gardner, why don't you offer to hire him for, say, six months? I'm sure he'll be glad to help you for a period of time."

"That might be possible. I will contact my superiors as soon as possible."

Chapter 9

Boris walks to the back of the plane, takes his cell phone out and dials. "Igor, it is Boris. I have the woman, Ms. Walters. She questioned, why don't we hire Mr. Gardner for a period of time, say six months? She thinks he will be glad to come."

"Boris, how can we hire him? We do not have any money. What are you thinking?"

"Igor, if we do not fix this rocket problem, we will be fired, sent to Siberia, and who knows what else."

"Boris, you forgot being hung by our finger tips."

"Ha, ha. Igor this is **not** funny. I am going ahead and call him and tell him we want to hire him. I will get a visa for him to enter Kazakhstan and give it to him as he enters the airport at Astana. I will tell him that Ms. Walters has agreed to work for us. Then he will surely come."

"Who buys his airline ticket?"

"He will buy his own airline ticket and, if he asks, I will tell him I will refund the cost when he gets here."

"This is sounding like some deep **doo-doo** to me, Boris. Call me when you get here with Ms. Walters."

"I will. Goodbye, Igor."

"Goodbye, Boris."

Chapter 10

Dialing, "Hello, Mr. Gardner, you don't know me but my name is Boris and I am from Kazakhstan"

"I have been waiting for your call." His voice rising, "You have taken my Director of Personnel, Jenna Walters, and I want her returned immediately."

"It won't be that easy, Mr. Gardner. You see, she has consented to come to work for the Russian space effort."

Indignantly, "Not willingly, she didn't. I know that for sure."

"Well, willingly or not, she has. And, if you want her back unharmed, you will have to do the same."

"You are kidding! Are you saying I have to come and work **for you?**"

"Yes, that is right, for six to eight months, until we find the problem in our Soyuz rocket and get it back on track."

"You say six to eight months.?"

"Yes."

"I will not do anything to harm the United States and their space effort. Is that clear?"

"Perfectly."

"I want to speak with Ms. Walters now."

"Just a moment. Here she is."

"Yes, Phillip, I'm here. It's so good to hear your voice."

"And yours, too. Have you told them that you'll work for them for 6 to 8 months?"

"Yes. I didn't think I had any other choice."

"Does the Russian government know anything about this? It looks like a kidnaping to me, plain and simple."

"I think so, too. Here, Boris wants his phone back. Goodbye, Phillip."

"Goodbye, Jenna."

"Mr. Gardner, if you have a pen and paper I will give you my cell phone number where you can reach me. Also, there is a flight on Trans Aero Airlines that you can take from Miami, with a couple of connections, to Astana, Kazakhstan, our capital. I will meet you there at the airport with a visa and Ms. Walters. From there we will drive south to the Baikonur Cosmodrome, which is the location of the Russian space effort and where I work with my brother."

"I'll call you as soon as I find out all the flight information."

"That will be good." They hang up.

After a few moments, Boris dials a number inside Kazakhstan, that of his supervisor at the Cosmodrome, Viktor Goosevick.

"Viktor, this is Boris."

"I know, I recognize your voice. We have missed you the last few days. How was your vacation?"

"It was very good. I contacted a friend of mine who owns a company in the U.S. called Space Flight and he has agreed to come here for a few months to help get our Soyuz rocket program back on track. His name is Phillip Gardner."

"Why, that is excellent, Boris! I have heard a lot about him. He is well-respected in the space flight industry. But, how much money does he expect to make?"

"Whatever you can find in the budget, plus a place for him and his assistant to stay and a company car."

"I think we can put a handsome package together for them. Is one villa with two bedrooms sufficient?"

"I believe that will be fine, plus two Identity Badges."

"Yes, yes, I know. I will prepare everything. When can we expect them…in about a week?"

"No, I think sooner, but I will let you know when I know more."

"That is excellent, Boris. I will wait for your call. Goodbye and good work!"

Chapter 11

Ms. Walters, we will be landing at the Astana International Airport in Kazakhstan in about twenty minutes. I have called ahead to order a visa for you to enter the country and it will be at the airport when we arrive."

"And what about Phillip Gardner?"

"He will be on his way here as soon as he can make arrangements, which I believe will be in a day or so. You and I will get two rooms in Astana near the airport until he arrives. That is, if I can trust you. Or, should I just get one room?"

"No, no. Two rooms are fine. You can trust me."

"Mr. Gardner has my cell phone number and I am sure he will call when his plans are firm."

Two or three minutes pass. "Boris."

"Yes, Ms. Walters."

"Do you believe there is a God?"

"A God?"

"Yes. Someone who has created the earth and everything and everyone on it?"

"Well, I have not given it much thought. I know there is an order; the sun and the moon shine when they are supposed to....Uh, I can tell we are starting down. Please fasten your seat belt for landing. And Ms. Walters."

"Yes."

"Please cooperate and let everything go smoothly."

"Yes, I will. I can tell God's will is unfolding and I want to flow with it."

After a minute or so, "Ms. Walters, as you can see out the window, we will be landing in an area of the country that is low-lying, though Kazakhstan has everything from that to snow-covered mountains.

"We are part of central Asia and, the part west of the Ural River is in Europe. Kazakhstan borders on Russia, China, Kyrgyzstan, Uzbekistan, and

Turkmenistan and also adjoins a large part of the Caspian Sea.

"Politically, we have shifted from nomadic tribes to Genghis Khan in the 13th century and back to nomadic tribes with their ensuing struggles and the tsars. The Russians began advancing into Kazakhstan in the 18th century and by the mid-19th century, all of Kazakhstan was part of the Russian Empire. We became a bi-lingual country, Kazakh and Russian. Following the 1917 Russian Revolution, ensuing civil wars, and Stalin's repression of our culture, Kazakhstan became the Kazakhstan Soviet Socialist Republic in 1936.

"Kazakhstan was the last of the Soviet Republics to declare independence following the dissolution of the Soviet Union in 1991. Russia now rents the Baikonur Cosmodrome from Kazakhstan, and will, until its lease ends in 2050.

"You asked me about a God. Seventy percent of our population is Islamic, and Christianity is practiced by twenty-six percent of our people. Kazakhstan allows for freedom of religion."

"And which are you, if I might ask?"

"My family background is Islamic, though I don't attend services regularly, only on some holidays....Here we are....touchdown. We have arrived."

"It's been a long time coming," Jenna remarks. "How many hours is the flight?"

"It can take twenty-four hours or so with layovers and delays due to unforeseen circumstances."

As they taxi to the terminal, "Boris, you sound very proud of your history. I hope you will tell this to Phillip when he arrives."

"I will when the time is right. Perhaps during the long drive from Astana to the Baikonur Cosmodrome."

"Yes. That sounds good."

"After we get through Customs, I will pick up the rental car left for us by my supervisor and we will head for a nearby hotel."

"That's with two rooms, right?"

"Yes, Ms. Walters."

"Ms. Walters, back on the plane, you mentioned, 'God's will'. What is that? How can I find God's will for me?"

"Boris, God has a plan for each of our lives, even down to the minutest detail. The plan He has for each of us is also known as His will for us."

"Ms. Walters, here is our hotel. Is it possible to talk with you again some time on this subject?"

"Absolutely, Boris."

"By the way, Ms. Walters, when my supervisor asked me if one apartment with two bedrooms was sufficient for you and Mr. Gardner, I said, 'Yes.' Is that O.K.?"

"Boris, actually, I hardly know Mr. Gardner. We've talked maybe three times. Other than what I read in his company brochure, I don't know anything about him."

"What do you want me to do?"

"I think I'd feel safer living in an apartment with Phillip than in an apartment by myself. Being kidnaped once was enough for me."

"Alright, I understand."

Boris' cell phone rings. "Hello, this is Boris. Ah, Mr. Gardner, it is good to hear from you. Yes, Ms. Walters is here. I will let you speak with her."

"Hello, Phillip?"

"Yes, Jenna? Are you O.K.?"

"Yes, I'm fine. When will you be here?"

"Tomorrow morning. I'll give Boris the flight number and arrival time."

"Phillip?"

"Yes."

"Are you married?"

"Well, no. Why do you ask?"

"Because Boris' supervisor is getting us one two-bedroom apartment. When Boris told me, I figured if I was by myself in a one-bedroom apartment, I could be kidnaped again. So I said go ahead and get one apartment with two bedrooms. Is that O.K.?"

"Yes, that's fine. We can look out for each other."

"Here's Boris," handing Boris the phone.

"Mr. Gardner, I have a pen and paper ready. Tell me your airline, the flight number, and arrival time at the Astana International Airport. O.K....Yes....We'll meet you tomorrow morning at 8:00am. Your visa will be at Airport Arrivals. Fine, we'll see you tomorrow morning. Goodbye."

"O.K., Ms. Walters, we need to get a good night's sleep. We will be checking out early, maybe 6:00am, and eating breakfast at the airport."

"Boris, I'll see you bright and early tomorrow morning, packed and ready to go."

Chapter 12

S o," driving away, "check-out was smooth and uneventful."

"Boris, you expected something else?"

"No, I am just thankful when things go as planned. We will be at the airport in about 20 minutes, which will be plenty of time for our breakfast and to meet Mr. Gardner as he arrives."

Not much is said on the drive to the airport. They park and walk toward Trans Aero Airlines. They show their passports to the guard at the door and enter.

"Ms. Walters," pointing to a café, "there is a place where we can have breakfast. Before we leave, let us order a roll and coffee for Mr. Gardner in case he is hungry."

"That's a great idea. He may not have eaten anything for a while."

They eat and exit the café, carrying the take-out order. An announcement is made over the airport loud speaker in Russian, Kazakh, and English saying that a Trans Aero flight was landing.

"Is that Philip's flight?"

"Yes, it is. After they taxi over here, it shouldn't be too long before we see him." The plane finally stops.

A pick-up truck, with a ladder mounted in a slanted-fashion on top, drives up to the airplane cabin door and the door soon swings open. After a couple of minutes, a flight attendant exits the plane first and walks down the stairs followed by passengers, one after another.

"There he is! There's Phillip!.... Phillip, Phillip," she calls out! He sees her and waves. They both begin to walk faster toward each other.

When Jenna reaches Phillip, he envelops her in his arms and says, "I didn't know if I'd ever see you again. I am so sorry this all happened to you. I'll be keeping a close eye on you from now on. Is that O.K.?"

Pulling her head back and gazing up into his eyes, "That will be totally fine, Phillip."

They smile at each other and he hugs her close again. After a few seconds, Boris clears his throat. Jenna pulls back and says, "Phillip, I'd like you to meet Boris. Boris, this is Phillip Gardner."

"My pleasure, Mr. Gardner," shaking his hand.

"Boris, we all know that Jenna and I are here because you kidnaped her and brought her here. And we have surmised that the Russian government knows nothing about this."

"Mr. Gardner, you are right. Neither the government nor my supervisor knows what I did. I told my supervisor that I had known you for some time and that I asked you to come here and you and your assistant accepted out of respect for our friendship."

"Boris....you **did** tell your supervisor that we would only be here for 6 or 8 months, didn't you?"

"Yes, that is the absolute truth, I swear." Boris raises his right hand and bows his head, "I very much regret what I did but please, please don't tell anyone or my brother and I will be killed."

"I remember your saying something about a brother."

"He is Igor and he works with me at the Baikonur Cosmodrome. We both worked on the third stage of the Soyuz rocket and we believe this is where the failure occurred on the May flight to replenish the International Space Station."

"Boy, this is a lot to take in. What happens now?"

"The three of us drive south to the Baikonur Cosmodrome. It's about a 7-hour drive, so we have to get started. I called my supervisor, Viktor Goosevick, last night and he has made arrangements for you to have a two-bedroom villa, a car and a salary. He has heard of your work and your company and is extremely happy that you and your assistant consented to come."

Jenna adds, "And not knowing when you last ate, we brought you a roll and coffee." She hands Phillip the bag of food.

"Thanks, you both are trying to get on my good side."

"You've got that right," adds Jenna.

They arrive at Boris' car and Boris opens the back door. "Mr. Gardner, you get the back seat to yourself. You can stretch out and take a nap if you would like."

"I would like that, thanks. It's been a long 24-hours."

They all get situated and begin to head south toward the Cosmodrome. Phillip eats his roll, sips his coffee, and soon is making himself comfortable in the back seat.

Fifteen or twenty minutes pass and Boris remarks, "Ms. Walters, will you tell me more about your God?"

"Well, I think it's best if I start at the beginning, **my** beginning. At that time, a neighbor had called me a couple of times, asking me if I wanted to go to a Wednesday night service at her church. Each time she called, I had an excuse for why I couldn't go. Then she stopped calling. As the next Wednesday night approached and she hadn't called, I began to have a strong inclination inside of me to go to the church that evening. I called her and she said, "Great, be over here at 6:30pm and we'll go together."

"The evening began with joyful singing and the clapping of hands to the beat of the music. After about fifteen minutes of singing, the minister got up in front of the church and introduced the evening's speaker.

"The speaker was an older man, about my father's age. He said he had been a former city attorney, like my father, and an alcoholic, also like my father. But, unlike my father, he eventually saw what alcohol was doing to him and his family and he quit drinking. When the speaker asked if anyone in the congregation wanted to begin a new life, with the healing touch of Jesus walking with them, I didn't move a muscle.

"I think I was in shock, in shock that the God of the universe had impelled me to go there that night to hear a story from a man whose life paralleled my own father's life. And that was just the beginning. Later that night, when I was alone, I asked Jesus to come into my heart to be my Savior and Lord. It was a seemingly simple action on my part but it began an on-going process of forgiveness and change in me.

"Just because I had forgotten an event, did not mean it had gone away. I unknowingly would bury it deep inside me. Like a garbage can holds garbage, so too, had I been a receptacle for anger and unforgiveness. Holding anger and unforgiveness inside me was really, holding onto darkness. And darkness kills. I was the one being hurt by letting darkness stay inside me.

"Every time a thought came to my mind about someone, I knew that was God wanting me to forgive them and I did, sometimes with great earnest. The result? I've been changed into the person I am today, someone who is at peace with herself and with those around her. And it all began when I asked Jesus into my heart.

"I believe everyone has anger, bitterness, etc., in varying degrees, pushed down inside them, buried and forgotten. But it's not gone. Each person needs to ask God to reveal to them any unforgiveness in them and He will. Then, in no uncertain terms, tell the anger and unforgiveness to leave you and proclaim your forgiveness for that person. I can't help but think, that the person I forgave is, in some way, a little bit freer.

Chapter 13

Ms. Walters, that was staggering. Anytime you want to tell me more about your God, I am ready to listen."

"Boris, I'm kind of all talked out right now. I think I'll take a little nap." She leans over to the right and against the window. After several minutes she is out.

Boris takes out his phone and dials. "Igor, it is Boris. Yes, they are both in my car sleeping and we are headed south toward the Cosmodrome. We hope to see you tomorrow at work. We might be a little late, due to the time change. Alright, I'll see you tomorrow. Goodbye."

The hours pass by quickly and an hour or so away from their destination, Boris takes out his phone and dials his supervisor, Viktor Goosevick. "Viktor, it is Boris. We will be there in about an hour. Is their apartment in the same complex as mine?"

"Yes Boris, it is in the building just west from where you live. You can get keys for their villa and for their car, which should be parked in front of the villa, from the building manager when you arrive. He will direct you to it. I will be at their apartment around 10:30am tomorrow to pick up Mr. Gardner and his assistant. You can follow us in your car."

"Yes, Viktor. Goodbye."

Chapter 14

Another half hour passes and Boris decides to wake up his passengers. "Hello, sleepers, it is time to wake up. We will soon be arriving at your villa"

"Oh, wow." rubbing her neck, "How long have we been sleeping?"

"A few hours. You must have needed it."

"Phillip, are you awake?"

"H-m-m, almost."

"I live in the same complex as you. Your villa is in a building just west of mine….Here is the gate and I will enter the numbers required to open it, 5, 7, 8, 2. I will pull up to the manager's office and get the keys to your villa and to your car, which, I believe, is parked in front of your villa. I will be right back."

"Well, Phillip, we're almost home."

"I think you're right, Jenna. Here comes Boris."

"All right," getting into the car, "I am going to drive you to your villa. First, we will pass my villa. And there it is on the right. Now, a little further into the next block is your villa, number 460 Borcht Lane. Over here is your car," getting out and handing Phillip both sets of keys. "Let me help you with your luggage. There is a small grocery store, out the entrance gate and to the right about a half a mile on the right. The money we use is the tenge. Two hundred forty tenge are equal to $1.00 US. I am giving you $50 worth or 12,000 tenge. You will be given more tomorrow at the Cosmodrome. But this will be enough for you to get things for breakfast tomorrow.

"Viktor Goosevick, my supervisor, will be coming to pick you up tomorrow morning at 10:30. I will follow you in my car and will bring you home in the evening."

"Should we call you when Viktor arrives?"

"Yes, I think that will be good."

Boris enters his car and drives back to his villa.

"Ms. Walters, shall we look at our home for the next few months?"

"I can't wait." They step in and view inside the kitchen and living room then, make their way down the hallway. Both bedrooms are similar.

Phillip says, "Ladies first. You pick which one you like, and I'll take the other."

"Well, I guess I'll take the first one. I'm keeping you as a buffer between me and the back door."

"Well then, I think I'll take the second one. I'll bring the luggage in. Why don't you shower first and I'll go to the store down the road?"

"Wait. I want to go to the store with you. I don't want to miss anything. I'll be ready in a couple of minutes," as she heads for the bathroom.

"I think I'll follow you in there. It's been a long ride in the car."

After a couple of minutes, "O.K. Jenna, let's go shopping." They enter the car, fasten their seat belts, and they're off. "Boris said after the gate, turn right and go about a half mile and it's on the right. Let's see if we can find it."

"I think I see a small store ahead on the right. This should be an adventure. I wouldn't miss this for anything."

Phillip parks and they exit the car.

"Let's see if they speak English," walking in. "Hello, do you speak English?"

"A little," says the lady at the cash register.

"We would like cereal, milk, coffee, sugar, sweet rolls, butter and dish detergent with a brush," handing her his list.

"Yes, please get a carrying case and follow me." In a matter of minutes their case is full and they are ready to check out.

"That will be 2,880 tenges."

"Alright." Phillip counts out 3,000 tenges and tells the cashier to keep the change.

As they walk out, Phillip says to Jenna, "I felt like I was counting out monopoly money."

"I know, it sure looks like monopoly money."

Soon they are back at their villa, placing the things in their refrigerator. "Jenna, why don't you use the bathroom first."

"Sounds good to me," as she heads for her bedroom to gather things for her bath. After about ten minutes, she exits the bathroom and says to Phillip who's sitting in the living room flipping through the TV channels, "The bathroom is yours."

"O.K., Thanks, I've been trying to find an English-speaking channel but so far, no luck. It would be nice to know what's going on in the rest of the world."

"I'm sure there's an English-speaking channel somewhere on there. I'm going to bed to get some rest before tomorrow. By the way, what time is it here now?"

"Well, let's see. It's ten hours ahead of Florida so that would make it….9:00pm."

"I have an alarm clock. Should I set it for 9:00am?"

"That should give us enough time to get dressed and have breakfast."

"O.K. I'll see you tomorrow. Goodnight, Phillip."

"Good night, Jenna."

Jenna heads for her room, finishes unpacking and soon, is out cold.

Chapter 15

The alarm crowed as loud as a rooster and it didn't take long for Jenna to get out of bed.

"I'd better knock on Phillip's door to wake him. Phillip, knock, knock. It's 9:00am....time to get up."

"O.K., thanks, I'll be there in a little bit."

By 9:45am, they are both dressed and busy in the kitchen making their breakfast. Soon they are sitting across from each other and Jenna asks, "Would you like to thank God?"

"Yes, I would....Oh, God, we are so thankful for our safe arrival here in Kazakhstan, for a good night's sleep and now, for our food. Please bless our day and may we be a blessing to the new people we meet today. Thank you, God. Amen."

"That was really nice, Phillip."

"Thanks, I really hope we are a blessing to those people we meet at the Cosmodrome. Would you like your sweet roll heated?"

"M-m-m-m, please."

"Two warmed sweet rolls coming up."

As 10:30 was approaching, they were finishing breakfast.

A knock on the door signaled Viktor's arrival. Phillip quickly called Boris and told him Viktor was at the door.

"Thank you for calling me, Phillip. I will see you at the Cosmodrome later."

"Jenna, are you ready; Viktor is at the door."

"Yes, I'm coming."

Soon, they were both in the car with Viktor, heading for the Cosmodrome.

"Viktor, how long is the ride from here to the Cosmodrome?"

"About twenty minutes, but by the time we park and walk in, it's about thirty minutes. The first thing we will do when we get there is get you both photo I.D. badges and a map of the area, both inside and outside the Cosmodrome. When you are finished there, I want you to get with Boris so he can show you what he does. If you see something that could have gone wrong with our last flight in May, tell Boris. He thinks the failure might have originated in the third stage which he and his brother have worked on."

"I will do that. Where will I find him?"

"After I show you to your desk, you can reach him by dialing x-468 on your desk phone. I will have an extra chair brought in for your assistant. Here we are, your desk for the next six months. I will let you both get to work. And, thank you for coming."

Chapter 16

After they are both seated, Phillip dials x-468. "Boris, it's Phillip Gardner and Jenna Walters and we are at x-314."

"Good, I will be there in two or three minutes."

As they wait for Boris, they go through the desk drawers to see what kind of supplies they have to work with.

Jenna remarks, "The desk seems pretty complete with all the things we might need. Here comes Boris."

"Ah, Mr. Gardner and Ms. Walters, my travel partners," shaking their hand, "I see you have both gotten your identification badges. Now I would like to give you a little background of our Russian Space Program and the Soyuz rocket," as he pulls up a chair.

"Our Soyuz rocket and spacecraft consists of an Orbital Module, a Descent Module, and an Instrumentation/ Propulsion Module.

"The Orbital Module is used by the crew while in orbit during free-flight. There is a hatch and rendezvous antenna located on the front end. A docking mechanism is used to dock with the space station and the hatch allows entry into the station.

"The Descent Module is where the cosmonauts and astronauts sit for launch, re-entry, and landing. All necessary controls and displays of the Soyuz are located here.

"You will find the next part interesting. This module also contains custom-filled seat liners for each crew member's couch/seat which are individually molded to fit each person's body. This ensures a tight, comfortable fit when the module lands on earth and it has to withstand a bumpy landing. The space craft is rigorously tested, first on the ground, in hanger drop landing tests, in air drop tests and in space before it is declared flight-ready. When crew members are brought to the station aboard the Soyuz shuttle, their seat liners are brought with them and transferred to the space station as part of the crew handover activities. The seat liners are transferred back to the Descent Module when cosmonauts are ready to return to earth. The Descent Module, with its heat shield, is the only portion of the Soyuz that survives the return to earth.

"The third module is the Instrumentation /Propulsion module. The Propulsion compartment contains the system that is used to perform any maneuvers while in orbit and also, to perform the deorbit burns necessary to slow down the module as it returns to earth and then, it is jettisoned.

"The rendezvous and docking with the space station are both automated, meaning once the spacecraft is within 492 feet of the station, the Russian Mission Control Center located just outside Moscow, monitors and controls the approach and docking with the station though, the Soyuz crew has the capability to manually intervene and execute these operations.

"Most of our flights are to the International Space Station, the ISS. Each launch costs approximately 70 million Euros or $79,000,000.

"There are several versions of the Soyuz rocket, a Soyuz-2.1a was used in May. This rocket, like all the Soyuz rockets, was assembled horizontally and during that time, the Progress cargo spacecraft was bolted onto the top of it. Together, they were placed on a carrier, which is mounted on train rails and then rolled out to its launch pad. The three stage rocket with spacecraft was lifted from its horizontal position on the train rails to vertical position and onto the launch pad by a hydraulic lift system. Rotating platforms were then moved around the rocket to give workers access to the vehicle during final flight preparations.

"A sample of some of the supplies loaded on a cargo flight are:

1,146 pounds of propellant to refuel the space station's service module,

926 pounds of water,

105 pounds of air and oxygen,

3,071 pounds of dry cargo including:

- Food, napkins, food waste containers,
- Toilet devices, liners and waste containers,
- Medical supplies, personal hygiene items and air monitors.
- Batteries, computers, chargers, tools and fire extinguishers.
- Science equipment hardware
- American and European equipment.

"Normally, the Soyuz-2.1a rocket reaches orbit about ten minutes after launch. It circles earth four times while various checks are performed. During that time, the Progress spacecraft will deploy its power generating solar panel wings and navigation antenna. Then they will begin thrusting for a two-day rendezvous with the space station. The recent unmanned flight was loaded with three tons of equipment and supplies for the ISS. However, there was a glitch in one of the stages and the Progress cargo ship was stranded in an orbit too low to reach the station. The rocket could not overcome the force of gravity pulling down on it at that altitude and nine days later, the capsule fell back into Earth's atmosphere and was incinerated. So that is where we are now; needing to find an answer to this flight's failure. "Now," getting up, "I would like us to take a short walk to the video room. You might want to bring a pad and pencil to take notes."

As they walk, "What will we be seeing, Boris?"

"We will be watching a video of our Soyuz-2.1a rocket lifting off in May of this year. The first two stages seemed to go well, but when the third stage should have ignited, it didn't.

"So please sit down and I will signal the video person to start it."

As the video ran, Phillip was making notes. "Boris, is it possible to run it again slower and I want to stop it in several places?"

"Just a minute." Boris picks up his phone, calls the video person in the booth, and says something to him in Russian. "Alright, it will be played in half speed. When you want it stopped, raise your left hand. When you want it continued, move your right hand in a circle motion."

Phillip moved his right hand in a circle and the video began again, this time in slow-motion. In the next hour or so, he stopped and started it several times, making notes each time it stopped. Then Phillip questioned Boris, "Do you have a mock-up of the rocket? I'd like to see it for myself."

"Yes, we do but on the way there, why don't we stop at the cafeteria for some lunch before it closes?"

"Oh, what time is it?"

"It is now 12:40pm and the cafeteria closes at 1:00pm. The hours it is open are posted at the entrance of the cafeteria. Here we are," motioning toward the entrance. "The Cosmodrome cafeteria is open twenty-four hours a day, seven days a week. Hot meals are served three hours each day: 8-9am, 12-1pm, and 5-6pm. Sandwiches, snacks and drinks are available twenty-four hours a day. When you are given your check, print and sign your name on the back, write your telephone extension next and then, drop it in the box in front when you leave."

"I guess we have a lot to learn. Are you taking this all in Jenna, in case I miss something?"

"Yes, Phillip, between the two of us, I think we'll get it all."

Lunch didn't take long and when they finished, they headed toward the Soyuz mock-up rocket, similar to the one that self-destructed in May.

"Here we are! The majesty of it always amazes me. Mr. Gardner, what do your notes say about it?"

"Boris, you were right in saying that we need to check the connection between the second and third stages but that might not be the only problem. I need to study the schematics of the connection. Could I see them?"

"Of course, let us take a seat in the control room and I will pull them up for you. There are several to look at."

"Let's see them in the order that they fire. First, the first stage as it readies to fire."

"Here we are. Each stage follows the same sequence of firing." For the next two hours, Phillip and Boris study the schematics with a proverbial fine-toothed comb.

As the clock approaches 5:00pm, Boris says, "We have something to do before we leave today. And that is to pay you in advance for the next few weeks."

"M-m-m, I'd like that. I'm ready to go."

"Let us go to the Pay Department and then I will drive you both home."

As they drive, "Boris, I'd like to take Ms. Walters to a nice restaurant for dinner this evening. What would you recommend?"

"Do you like Kazakh food?"

Looking back at Jenna, who nods yes, "Yes."

"I have liked the Kazakh Restaurant. Their steak is some of the finest. And it is not too far from where you live."

Boris gives Phillip the instructions on how to find the restaurant, and drops them off at their villa. "Now, will you be driving to the Cosmodrome by yourselves tomorrow morning?"

"Yes, what time should we get there?"

"Nine-thirty sounds good. Call me when you get there. I will see you both tomorrow."

What a nice, cozy restaurant this is," remarked Jenna, "and the waiter speaks a little English."

"Jenna, while you were getting ready for dinner, I found an English-speaking TV station and the newscaster mentioned that the Russian President had ordered troops with air support into Syria, ostensibly to back up the Syrian President, probably against ISIS. We might have gotten ourselves into something here. Right now, I think the only place we can talk about this is here, in a public place."

"Phillip, don't forget, I'm a former P.I. I'll check our villa for any surveillance devices when we get back."

"You'll probably have to do that every evening when we return from the Cosmodrome."

"That's fine. You drive home and I'll check for surveillance devices when we get there. My work won't take as long as your driving does."

"How is your steak, Jenna?"

"Absolutely, the most delicious I've ever had."

"Mine, too. Boris knows his meat. One evening we'll try Russian food. But back to what I saw on T.V., I'll be getting their Soyuz rocket going in less than six months. But while we're here, let's keep our eyes and ears open to anything, especially for any rocket that is being readied to carry a warhead aimed at the U.S."

They finish their meal and leave the restaurant, with Phillip taking Jenna's hand as they head toward the car.

After entering their villa, Jenna makes a sweep of each room, looking for any surveillance devices and the rooms were clean. She also checks out their car, with the same result.

"Jenna, as a P.I., what kinds of things do you think we should be looking and listening for at the Cosmodrome?"

"When we get to know their routine, I think we should be open to any change in that routine. When any new project comes up or any enhancements are made to the Soyuz rocket, let's remember them."

"I don't think we should make any obvious notes until we get back to the villa."

"Yes, Phillip, but remember, Boris sees you taking notes all the time, so he wouldn't think anything is unusual when you do take notes."

After a few moments, "I'm going to call Alex Wheeler at Space Flight and keep him appraised." Dialing his phone, "Hello Alex, it's Phillip Gardner. I'm calling to get you up-to-date on what's happening here. We're staying in a villa which is a twenty-minute drive to work at the Baikonur Cosmodrome in Kazakhstan. I've started working on evaluating their Soyuz 2.1a rocket. An unmanned version of it malfunctioned in May. It fell back into the atmosphere and was incinerated. I'm working on the 'why.' On another subject, earlier this evening, I heard on TV about Russia sending troops and air support to Syria. While I could finish my work here earlier, Jenna and I have agreed to take our time in order to watch and listen for any indication that Russia is aiming a rocket toward the U.S. Will you contact the F.B.I. at the Cape to inform them of this? They are familiar with Jenna being kidnapped and my following her here."

"Oh, wow, Phillip. You've got a lot going on. Just be careful."

"We will. How is Space Flight doing?"

"We're keeping to our schedule and everything seems to be going without a hitch."

"That's great. I'll try to call you from time-to-time to update you. Keep up the good work."

"O.K., I will. Goodbye, Phillip."

"Goodbye, Alex."

Turning his phone off, he turns to Jenna, "That was good news, Space Flight is on schedule. Of course," walking towards her, "I don't know how they can continue without your advice as Director of Personnel."

"Well, they'll just have to make do at this point."

He moves closer and encircles her with his arms, "We make a pretty good team, don't you think?"

"M-m-m, so far," looking up to him, "I think so."

He kisses her forehead and then opens his arms to let her go. "The bathroom is yours first."

"O.K., I won't be long."

Chapter 18

The next several days and weeks Phillip and Jenna continue to go to work at the Cosmodrome.

Each evening they review what they heard and saw during the day. One evening, Jenna said she saw a word printed on some boxes: "HEMP." Phillip knew right away it meant: "**H**igh Altitude **E**lectro **M**agnetic **P**ulse. It's meant to destroy power grids from some ten miles up in the sky. The bottom line is: any and all electronic devices will stop working. Cars will all stop, airplanes will fall out of the sky, power plants, telephone exchanges, cell towers, all would be dead. It destroys the functions of the transistors in electric circuits. It has little or no effect on living tissue. But it can lead to chaos in the streets when people's way of life is changed or destroyed."

"Phillip, is there any way to combat it?"

"Yes, but I have to tell Alex to alert the F.B.I. I'll call him immediately." Phillip dials his phone, "Alex, how is everything going? Good, good. Now, what I am about to tell you is very important. Get a pencil and paper....I want you to alert the F.B.I. that Jenna saw some boxes at the

71

Cosmodrome today that were labeled: **'HEMP'**. Those letters stand for **H**igh Altitude **E**lectro **M**agnetic **P**ulse. I want you to write this down and give it to the F.B.I. We need to be preparing the nation's electrical grid for an attack. Thanks, Alex. I'll be talking to you." He turns his phone off.

"Well, Jenna," turning to her, "we found what we were supposed find by working in the Cosmodrome. We can wrap up the work on the Soyuz rocket and leave. Though I really hate to leave here or I should say, I hate to leave my time here with you."

"I feel the same Phillip. It's been exciting and fun. We make a pretty good team, don't you think?"

"Yes, I do. It doesn't have to end here, you know."

"Oh?"

Phillip's phone rings. "I wonder who that could be. Hello?"

"Phillip, it's Alex. I just hung up from talking with the F.B.I., and they wanted me to ask you to stay there a little bit longer."

"Well, if the emphasis is on the words, 'little bit.' I guess we can but I'll have to check this out with Jenna. Hold on."

"Jenna, the F.B.I. wants us to stay here a little bit longer. What do you think?"

"I," looking down...."I guess so."

"Alex, she hesitatingly said, yes. So remember the emphasis is on the words, 'little bit'."

"Yes Phillip, I'll tell them. Goodbye."

"Goodbye, Alex."

Phillip turns off his phone and turns to Jenna. He opens his arms towards her. She walks into them and is gently surrounded and held. He hears a small sob coming from her. He begins to rock her and says, "It's O.K., it's O.K. Believe me, we're not staying a second longer than necessary."

She looks up at him and says, "Is that a promise?"

"Yes, cross my heart." He holds her close for another few seconds and then releases her. "Are you ready to call it a day?"

"No, not quite. Let's watch some TV for a few minutes."

They both sit on the couch and Phillip switches on the TV. Then he puts his arm around her and she snuggles up close to him. It doesn't take long for her to fall asleep leaning against him. He puts one arm under her arms and the other under her knees and gently picks her up and takes her to her bedroom. He lays her down, takes her shoes off, and covers her with a blanket. He stands there for a moment, gazing at her, and then turns to exit the room. He heads for his bedroom and is fast asleep in no time.

The next morning they are both awake and beginning to make breakfast. "Phillip, I'm sorry I fell apart last night. Did you carry me to bed?"

"Yes, and took your shoes off and covered you with a blanket."

"Thank you. I appreciate that. I think it all began to hit me, how dangerous this all is."

"Now, let's keep in mind that all we're doing is keeping our eyes and ears open. You were pretty good at spotting that HEMP. That's all we're doing, nothing more."

"Thanks for reminding me."

"Sure…Are we almost finished breakfast?"

"M-m-m, just about," finishing her coffee.

"O.K. then, let's get to work!"

As they are driving, "Why don't I ask Boris if we can take a long weekend off and drive to the mountains? How does that sound?"

"That sounds great! Ask Boris to recommend a place to stay."

"You're right, I'll do that."

The day progresses as usual. During the lunch break, Phillip asks Boris about taking a four-day weekend and going to the mountains.

"Mr. Gardner, I hear many favorable comments about the Ural Mountains. Many people who work at the Cosmodrome go there."

"Can you give me instructions on how to drive there? How is the weather there this time of year?"

"The Fall season is beginning so you will see the leaves changing color. The temperature can get down to the 30's at night so you will need some winter clothing."

"It sounds exciting, doesn't it, Jenna?"

"I'll say. And Boris could you give Phillip directions to find winter clothes that are not too expensive?"

"Yes, I know of a shopping center not too far from where you live. I will give Mr. Gardner driving instructions for that, too."

"I thought that a change of scenery would be good for us."

"Mr. Gardner, here are the driving instructions for both places."

"Can you get me a phone number so I can make reservations?"

"Yes, I will contact my friends and get a phone number for the place they stay."

"Boris, we'll be back at our desk in a few minutes."

"Alright, I will see you there."

Phillip leans forward and takes Jenna's hand. "I think we both needed some time away from here."

"We never get to see Fall leaves in Florida. I am so looking forward to this."

"Me, too. We'll go shopping for winter cloths one evening after dinner. In the meantime, we should get back to work."

The rest of the day goes smoothly and soon they are on their way back to their villa. When they're almost there, "Jenna, I think I'll call to make reservations for this weekend."

"Sounds good."

After a few minutes, "It's done. We have one double bed for a long weekend. Let's get a bite of dinner and go shopping."

"Now wait, just a minute." holding up her hand. "So, it's one double bed in one bedroom?"

"That's all they had left."

"Well, O. K., but you'd better not laugh at my P.J.'s.

"I won't, I swear. I'll just think of you as one of the guys."

"Fat chance of that. Let's go get some dinner." As they are walking, "You know, Phillip, I don't think God would approve of this."

"What, having dinner?"

"No," raising her voice, "**sleeping together!**"

"Well, maybe we could find a place to get married first. What do you think of that?"

"Phillip Gardner, are you asking me to marry you?"

"Yes." He kneels down on one knee, "My fair lady, will you be my bride?"

"Well, I'll have to think about it....Yes, my prince."

Chapter 19

Phillip stands up and they both embrace. Then Phillip says, "I'm going to call Boris to see if he can help us find a place to get married. Boris, it's Phillip Gardner. Jenna & I would like to be married before we go away this weekend. Do you know of a place nearby?"

"Well, we have a chapel in the Cosmodrome which you can use. I can find a Christian minister and someone to play the keyboard. How would that be?"

"Let me ask my bride. Jenna, Boris says there is a chapel in the Cosmodrome which we could use. He could also get a Christian minister and a keyboardist for music. What do you think of that? Would you like to be married in the Cosmodrome on Thursday afternoon?"

"I think that would be wonderful, and we could have an open invitation to anyone who wants to come?"

"O.K., Boris. Jenna says that's great, and we could have an open invitation to anyone who wants to come."

"I know Viktor will want to buy the wedding cake and champagne and vodka."

"Great. What time should it be?"

"What about 2:00pm? That would leave a little time to celebrate afterward. Russians always like to celebrate and make toasts. We might even have a band for dancing. I will make all of the arrangements for you two lovebirds. And, by the way, I will ask Viktor to give you another day off, so you will not have to return to work until Wednesday. How is that?"

"That's fantastic. I can't wait to tell Jenna. Thanks Boris, we'll see you tomorrow at work."

"Jenna, our little wedding is growing into something we won't soon forget. It's turning into a Russian celebration to end all celebrations. We'll even have a band for dancing our first dance together."

"Oh my goodness, we'd better practice. I mean, with hundreds of people watching....No pressure or anything....But seriously, you know Phillip, it sounds like they've taken us under their wing as their little chicks."

"Well, Boris did call us love birds. I guess we'd better just enjoy it. Come here my little dancing chickadee," holding his arms out to her.

Jenna moves toward him and they begin dancing to imaginary music which Phillip soon begins to hum.

"Boy, you're really getting into this. We even have music to dance to."

"Only the best for my little lady. Now, a slow tune, which we will need to come closer to each other to dance," as he holds her closer.

Soon they are barely dancing, just barely moving together when Phillip, looking into her eyes says, "Jenna, I love you **so** much, I never thought I could be so happy."

She looks up at him and says, "I love you, too; like I've never loved anyone ever." And they kiss, and they kiss, and they hold each other.

Chapter 20

The next morning at the Cosmodrome, Boris is running around making plans for the wedding the next day. He barely notices Phillip and Jenna when they arrive. "Welcome, welcome my little love birds. As you can see, things are falling into place for tomorrow. Viktor is beside himself with the planning. He loves it."

"Boris, can I pay you something, at least for the minister and keyboardist?"

"No, No. We have everything under control."

"Jenna and I might leave a little early today to go shopping and get everything ready for the weekend."

"That is fine, Mr. Gardner. I will probably be busy getting ready for tomorrow. I don't think much work will be done anyway. So, you take the time you need."

"O.K., Boris."

Back at their desk, Phillip tells Jenna, "It seems Boris and everyone else is getting ready for our wedding tomorrow. Boris said we could leave early today to go shopping for our wedding and

82

our honeymoon. So, let's leave after lunch and go shopping. What do you think of that?"

"That's great! I was beginning to wonder when we'd have time to shop."

After lunch, they are on their way to the shopping center. Jenna has her mind set on a simple wedding dress and something warm to wear in the mountains. Phillip needs a conservative suit and tie and some warm clothes. So the next few hours are filled with things like trying on dresses and suits. They plan to meet in the middle of the shopping center around 5:00pm and have dinner somewhere.

Chapter 21

The next morning they are up bright and early and doing things like packing for the weekend and putting their wedding clothes on hangers to take to the Cosmodrome.

"Phillip, who could believe they were getting married in the Cosmodrome in Kazakhstan? I can hardly believe it. Can you?"

"Well, when you think about it, it **is** pretty far-fetched. No one will believe it back home. I'll call Alex to tell him before we leave for the Cosmodrome."

Soon, they have their car packed for the wedding and weekend and are on their way to the Cosmodrome. Phillip reaches over to take Jenna's hand. "What are you thinking about?"

"I was just thinking that soon I'll be married to the most handsome man in the world. Now, I would call that God's blessing, wouldn't you?"

"I would say it's God's blessing that soon **I** will be married to the most beautiful woman in the world, who is beautiful, both inside and out."

"Well, we are starting our wedding day - in agreement," squeezing Phillip's hand.

Chapter 22

As anticipated, not much work was done that Thursday morning. Everyone was giddy, smiling at Jenna and Phillip when they passed their desk. Lunch was about the same, with congratulations flowing like water down a stream.

Soon Jenna and Phillip were in the chapel, ready to say their wedding vows to each other. The chapel was overflowing, with many of them attending a Christian wedding for the first time.

After the wedding, the real celebrating began. As the band started playing, Phillip asked Jenna for the first dance and they looked really, really good, just like they had rehearsed. After a couple of minutes, Phillip motioned for others to join them and the floor filled up quickly.

After the first dance, Viktor gave glasses filled with champagne to Jenna and Phillip, lifted his glass, and proposed a toast for them to have a long and happy marriage with many children.

Jenna cut the first piece of wedding cake and fed it to Phillip and he did the same for her.

A lot of vodka toasts were made for the motherland, Russia. Soon there were men dancing in native Kazakh costumes. It was fascinating for Phillip and Jenna who had never seen Kazakh dancers. Jenna whispered to Phillip, "They must have knees of steel."

There was more dancing and more toasts and then, as the clock approached 5:00pm, Phillip thanked Viktor for the wedding and told him that he and Jenna were leaving for their honeymoon in the mountains. Viktor gave them both a hug and thanked them again for coming to work in Kazakhstan. "We feel you are a part of our family here at the Cosmodrome."

"We do, too," Phillip and Jenna answered in unison.

Chapter 23

After they changed out of their wedding clothes at their villa, the drive to the Ural Mountains was a new adventure, with Jenna reading the map and Phillip driving.

"Look, we're making a good team again, don't you think, Jenna?"

"Yes, I do. You know Phillip, I'm going to miss these people at the Cosmodrome when we have to leave."

"Well, it's all been a good experience, especially our getting to know each other better, being married, and now, going on our honeymoon."

"I kind of wish I hadn't seen those boxes marked, 'HEMP'. That was scary. It's hard to think of your friends, as your enemies, too."

"I know, but now we're leaving that behind," taking her hand, let's just enjoy our time together on our honeymoon."

After a couple of hours of driving, "It's beginning to feel a little cooler now. I'll get our jackets from the back seat. This is so exciting. And I see some of the leaves on the trees are beginning to change color."

"The sun is beginning to set now, but we'll see more leaves changing color as we go higher into the mountains."

"Phillip, this is so heavenly, I can barely take it all in."

When they arrive at the resort, they find a rustic setting. The sun was just about gone, but with its last glimmer, the colors of the leaves gave a tantalizing hint of what lay ahead the next day. Jenna was giddy with excitement thinking about the grandeur that surrounded and awaited them. She was waiting in the car while Phillip went into the office to register. When he got back in the car, Jenna took his hand and said, "Oh, Phillip, God has painted a living picture for us that we'll see tomorrow. It will be His gift to us, His way of blessing us and placing His hand of love and sanctification upon our marriage."

"I sense it too, and I am so thankful."

They lean toward each other and kiss.

The next few days were spent hiking in the woods, hiking in the woods and, more hiking in the woods. They soaked in hues of color the likes of which, they never knew existed.

Before they knew it, they were busy packing their suitcases, getting ready to head back to their villa and to work.

As they drive back, Phillip comments, "Jenna, I know exactly what made Boris' rocket fail in May. I'll be telling him soon after we get back to the Cosmodrome. When we get to the villa, you'd better search for any listening devices, especially since we've been gone a few days. I think we'd better not say anything to each other until you're finished."

They brought their luggage in and Jenna began her search. In no time, she found one attached under the kitchen table and another under a table in the living room. Checking behind the pictures in the living room, she found another. She thought to herself, "This is going to take longer than usual. I sure don't want to miss any." In the bedrooms, she found two more. In the bathroom, one was stuck behind a mirror. Back in the kitchen, one was mounted in one of the cabinets. She placed them in a bag and whispered for Phillip to look inside. He leaned over, peered in, and his eyes widened. Then he looked at Jenna questioningly.

Again whispering, she told him she was going to wrap them in several bags and put them in the garbage bin down the street. "Do you want to come with me?"

"Absolutely!" Outside, after walking a few steps, "Jenna, this gives us a new impetus to leave as soon as possible. Why do you think there were so many? Don't you think that was a little overkill?"

"I have no idea of their reasoning. They obviously didn't know what I did for a living. And also, they were very serious about knowing what we were up to and what we were planning."

Chapter 24

The next morning, they were both in the car and on their way to the Cosmodrome by 9:00am.

"So, Jenna, we'll see how the day goes after I tell Boris what was wrong with the Soyuz rocket that failed."

"Do you seriously think they might try to stop us from leaving Kazakhstan?"

"They might try to talk us out of leaving; I'd be surprised if they didn't. But their putting all those listening devices in our small two-bedroom villa was heavy-handed and threatening. It's not something a friend would do."

"I agree."

A few minutes later, they had arrived at the Cosmodrome and were riding the elevator to the third floor and their desk. Phillip was soon calling Boris, asking him to come with a notepad and pencil.

"Boris, good to see you," shaking his hand.

"Mr. and Mrs. Gardner, I hope you both enjoyed your days off?"

"Yes, we did. Thank you for asking. Please sit down, Boris. I've been doing some thinking about the Soyuz rocket that failed and I believe it was due to some fuel pipes that froze. Since the fuel wasn't flowing properly, the third stage couldn't ignite. It wasn't your or Igor's fault."

"This is something I never would have guessed, Mr. Gardner. Igor and I will go back to the drawing board and trace this."

"Boris, my new bride and I are anxious to get back to our home in Florida and begin our new life together. We are planning for this to be our last day here and are heading back home tomorrow."

"Mr. Gardner, this is so sudden. I know Viktor will want to talk with you."

"That's fine. Tell us what time to be in his office."

"I will call him to see what is a good time to meet and I will let you know." Phillip stood up and shook hands with Boris as he left.

Boris couldn't get to his desk fast enough to call Viktor and get an appointment time. "Viktor, it is Boris. I am calling to let you know that Mr. Gardner has found where the failure occurred in the May launch of the Soyuz rocket."

"That is wonderful Boris. His honeymoon must have helped him think better," chuckling.

"Viktor, he and his wife are leaving tomorrow to go back to Florida to begin their new life together. I told him I knew you would want to talk with him. What time should I tell them to meet with you?"

"As soon as possible. Tell them to be at my office at 10:30am. Thank you, Boris.

Dialing his phone, "Mr. Gardner, could you and Mrs. Gardner meet with Viktor at 10:30am?"

"Of course," looking at his watch, "we'll leave in a few minutes. Thanks, Boris. Well, Jenna, this will be the great time-of-trying-to-convince-us-to-stay."

"I'll let you do all the talking. You've done great so far."

Soon they were on their way to Viktor's office. They knock on the door and his secretary says, in English, "Please come in and go through that door."

"Ah, Mr. and Mrs. Gardner, it is so good to see you both," shaking their hand. "Please, sit down. Boris told me that you were thinking of leaving tomorrow. This is so sudden," holding his arms apart in a questioning gesture.

"I know it is, but I was able to find the problem in the May Soyuz rocket failure and my wife and I are anxious to begin our new life together in our home in Florida."

"I understand this but I just want to let you know that we will miss a man of your stature and brilliant thinking here and are willing to offer you a handsome salary to stay."

"I appreciate that, but…"

"How would you like $1,000,000 a year, plus a very nice villa and car? We would even give you a title; maybe, Director of Rocket Research?"

Phillip turns to Jenna and looks at her questioningly. "Viktor, we both love it here and appreciate the wedding you gave us, but we are ready to go home."

"You heard it from the boss," as Phillip stands up and reaches for Jenna's hand, "Jenna and I are leaving tomorrow to go back home and begin our life together." Extending his hand to Viktor, "Thank you for everything, Viktor." And they exit his office.

After a few steps Jenna questions, "Did I hear you say **I** was the boss?"

"No, no, you can wipe that from your memory." He takes her hand and says, "Let's go."

They ride the elevator down to the third floor and Phillip calls Boris to ask him to meet them at their desk.

"Boris," shaking his hand, "Please, sit down. I just wanted to tell you how much we have enjoyed working with you these past few months."

"And I with you."

"At first, I saw it as bad, with your kidnapping Jenna and all, but God turned it into good, as He promises to those who are His children. Now, what I'm going to say next, I don't want you to tell Viktor until after we are on the plane headed home.

"When we returned to our villa after our honeymoon, we found listening devices planted there. And not just one or two. There were seven listening devices. Jenna, who is a private investigator, found them in every room, even the bathroom. We both thought this was heavy-handed and even more, it was threatening. It was not the way one friend would treat another. That's the reason we are leaving earlier than we expected.

"Because you have been our friend, we wanted to let you know. But again, please don't tell Viktor until after we have taken off late Thursday afternoon."

"Thank you for telling me this. I will relish telling Viktor about the mistake he made. He so easily tells others about their mistakes."

"So, Boris, about our rental car, I'll return it to the rental car place near the airport. Is that right?"

"Yes, and tell them it was rented by Victor at the Cosmodrome. They will know what to do." Turning to Jenna, "Mrs. Gardner, I am so thankful for our talks about your God."

"I am, too."

They all stand up, give each other a hug, and say their goodbyes. Phillip and Jenna leave the Cosmodrome for the last time.

The next few hours were spent packing and getting ready to leave for Astana early the next morning.

Chapter 25

The ride to the Astana International Airport was one of happy anticipation, knowing they were headed to a place that was free and not threatening - their home. They had lunch at the same café that Boris and Jenna had visited several months earlier. And they bought a snack to take with them. A little after 1:00pm, they were boarding the airplane.

Only when they were up in the air did Phillip and Jenna breathe a sigh of relief. Jenna remarks, "I wonder what Boris is saying to Viktor?"

Back at the Cosmodrome, on the phone, "Viktor, I need to talk with you about Mr. & Mrs. Gardner."

"Come right up, Boris."

Boris takes the elevator up to the top floor, exits, walks the few steps to Viktor's door, and knocks. The receptionist, in Russian, tells Boris to come in and go into Viktor's office.

"Boris, it is good to see you. What is it you want to tell me about Mr. & Mrs. Gardner? They are gone, is that right?"

"Yes, but I met with them yesterday before they left the Cosmodrome. Did you know Mrs. Gardner is a private investigator in the U.S.?"

"No, go on."

"When Mr. & Mrs. Gardner returned to their villa from their honeymoon, Mrs. Gardner searched the villa for any listening devices and found not one or two, but seven devices, one in every room of their villa, including their bathroom.

"Viktor, do you know what their reaction was when they found so many devices?"

"No, I don't."

"Well, first of all, they wondered what you were listening for in their bathroom. It was all so intrusive that they became fearful.

"So Viktor, if you are wondering what caused them to leave so suddenly, it was because of all the listening devices you had planted in their villa. You are the one who caused them to leave, pure and simple."

"That is enough, Boris. No more. You have made your point."

Chapter 26

Phillip and Jenna both look out the window as their plane gains altitude. "It's such a beautiful country, isn't it Phillip?"

"Yes, it's too bad it had a 'Viktor' living in it or we might still be there."

"Phillip, how many times did we comment on what a great team we were?"

"I lost count after four or five. Why do you ask?"

"I was just wondering, what if the F.B.I. asked us to go on another job for them?"

"M-m-m, I think we'll cross that bridge when we come to it."

"Phillip, I'm just saying, don't be surprised if they call."

.

Thanks to: NASA for their information regarding the Russian rocketry program. NASA had no part in writing or publishing any part of this book.

Special Thanks to:

Lindsey Adams – for her proofreading and suggestions.

and

David Adams – for suggestions he made after reading various scientific magazines.